···ANOTHER EXTRAORDINARY ANIMAL···

ROSIE
THE RIBETER

WRITTEN BY
Darcy Pattison

ILLUSTRATED BY
Nathaniel Gold

The Celebrated Jumping Frog of Calaveras County

Rosie, the Ribeter: The Celebrated Jumping Frog of Calaveras County
by Darcy Pattison, illustrated by Nathaniel Gold
Text © 2019 Darcy Pattison
Illustrations © 2019 Mims House

For permissions write to:
Mims House
1309 Broadway
Little Rock, AR 72202

MimsHouse.com

The Jumping Frog Jubilee® is the trademarked name of the annual frog jumping contest held by the 39th District Agricultural Association, State Agency, California. This book is not affiliated with the Jumping Frog Jubilee®.

ACKNOWLEDGMENTS: Thanks to the support of Henrietta Guidici, Bill Guzules, Dennis Matasci and Henry Astley, Ph.D.

Publisher's Cataloging-in-Publication data

Names: Pattison, Darcy, author. | Gold, Nathaniel E., illustrator.
Title: Rosie the Ribeter : the celebrated jumping frog of Calaveras County / written by Darcy Pattison; illustrated by Nathaniel Gold.
Series: Another Extraordinary Animal
Description: Little Rock, AR: Mims House, 2019.
Summary: Rosie the Ribeter, a female bullfrog, sets the world record in the Calaveras County frog-jumping contest.
Identifiers: ISBN 978-1-62944-073-6 (Hardcover) | 978-1-62944-074-3 (pbk.) | 978-1-62944-075-0 (ebook) | LCCN 2017904689
Subjects: Frogs--Juvenile literature. | Amphibians--Juvenile literature. | Calaveras County (Calif.)--Juvenile literature. | BISAC JUVENILE NONFICTION / Animals / Reptiles & Amphibians
Classification: LCC QL668.E2 .P38 2018 | DDC 597.8/9--dc23

On a warm night in May 1986, a wild bullfrog crouched with her eyes just above water. The deep calls of other bullfrogs filled the air.

Nearby, three men pushed a flat-bottom boat into the irrigation ditch. Lee Giudici, Bill Guzules, and Dennis Mastasci called themselves the World Champion Frog Jumping Team.

It was time to hunt for frogs to enter into the Jumping Frog Jubilee® contest held in Angels Camp, Calaveras County, California.

They paddled down the irrigation ditch. As they caught frogs, they put them into a damp burlap sack.

Near the ditch's end, a large tree spread above the water. Lee turned his head, flashing his bright headlight back and forth. The light reflected from a bullfrog's eyes. Temporarily blinded, the bullfrog didn't move. Lee leaned over the boat's edge to snatch up the frog.

Lee glanced at the frog's ear, a circle on the side of its head near the eye. It was smaller than the eye. That meant it was female. Because it was breeding season, they seldom found a suitable female. Turning her upside down, Lee made sure she wasn't pregnant. There were no egg sacs. The contest said the frogs must be at least four inches long from nose tip to rump. She looked to be larger than that. Her belly felt flat. That meant she wasn't weighed down by a recent meal of a crawdad, a bullfrog's favorite food. She was just a plain female bullfrog. Excited, Lee put her into the burlap sack with the other frogs. Maybe she was the frog to win that year. The men turned around and continued to hunt and catch frogs as they paddled back to their truck. By now, the sacks were full of frogs.

The next day, the team tested each bullfrog's jumping ability. On Lee's driveway, they rolled out a long carpet. Each bullfrog was dropped onto a pad about the size of a dinner plate. The men, now called frog jockeys, leapt at the frog to encourage it to jump farther.

The jumping frog contest was based on the length of a triple-jump. The team tested each frog and measured the distance of three jumps from the starting pad. If a frog jumped a different direction on each jump, it might end up very close to the starting pad. Instead, the team looked for frogs that jumped straight. Frogs that jumped crooked were put into a sack to return to the ditch that night.

The female bullfrog jumped straight. Even more unusual, she jumped once and stopped. Most bullfrogs jumped several times in a row. When a frog paused between jumps, it let the frog jockey catch up.

She was even more special than they had hoped.

After several days of selecting frogs, the team had about
60 straight-jumping frogs.

Meanwhile, Lee's next-door neighbor, Mrs. Joanne Nash, was a schoolteacher. Each May, she asked her students to suggest unusual names for frogs. She gave the list of names to the frog jumping team to use for the contest.

At the Jumping Frog Jubilee®, Henrietta Giudici, Lee's wife, filled out the registrations for the team's frogs. Using Mrs. Nash's list, Henrietta randomly chose a name to write on each entry form. Now, the female bullfrog was called Rosie the Ribeter.

In the Preliminary Round of the contest, over 1000 frogs jumped. Only the top 50 frogs with the longest triple-jump distance would compete in the Championship Round.

When officials measured, Rosie the Ribeter had jumped about 19 feet. That was good enough to qualify for the Championship Round. But it was still about two feet short of Weird Harrold's 1984 world record of 21 feet 1 ½ inches.

The frog jockeys let Rosie rest in the frog hotel. They made sure she had the right temperature and enough moisture. Too cold, and she'd go into hibernation. Too hot, and she'd dry out.

At the Championship Round, Lee hopefully dropped Rosie on the starting pad. He leapt at her with his mouth wide open. She thrust her legs in a tremendous leap and stopped.

Her next jump was lined up with the first,
a straight line.

Lee lunged again. This time, Rosie took an enormous leap. And she was done.

The officials marked the spot where she landed and measured. Rosie the Ribeter had won!

Even better, she had broken the world record with a jump of 21 feet 5 and ¾ inches.

The team celebrated!

Later that night, Lee, Bill and Dennis returned Rosie the Ribeter to the deep water under her tree.

Lee said, "Nature is nature. She belongs back where she came from."

The plain female bullfrog was born wild and died wild. She lived her entire life wild, except for that week in May 1986 when she set a world triple-jump record.

Over 30 years later, the world is still amazed because that record has never been beaten.

She has earned the name,

Rosie the Ribeter,
the Celebrated Jumping Frog
of Calaveras County.

North American Bullfrog

Rana catesbeiana

North American Bullfrog is the largest true frog species found in North America. They can weigh up to 1.1 pounds (0.5 kg). Their average length is 4-6 inches (10-15 cm). Bullfrogs are usually shades of brown and green, with the darker colors on their backs. Their back feet are fully webbed. To tell the difference between a boy and girl bullfrog, look at its eardrum. Male eardrums are larger than his eyes, while female eardrums are the same size as her eye or smaller. Also, during the breeding season the throat of the male bullfrog is yellow, and the female's is white. The average bullfrog lives seven to nine years in the wild, and up to 16 years in captivity.

Bullfrogs are native to eastern North America. However, they are an invasive species in California, having been introduced there in the early 1900s.

Bullfrogs are known for their loud, deep bass call that is repeated several times. It's been described as "jug-o-rum" and "br-rum." From a distance, the call sounds like a bull roaring.

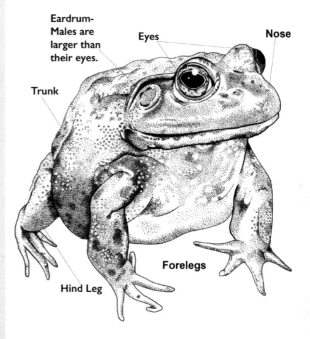

Eardrum- Males are larger than their eyes.

Eyes

Nose

Trunk

Forelegs

Hind Leg

Mark Twain, American Author

In January 1865, Samuel Langhorne Clemens (November 30, 1835 – April 21, 1910) was in California working as a miner. After two weeks of rain, he and his roommate decided to visit nearby Angels Camp in Calaveras County. For fun, the townspeople told tall tales. A tall tale is an exaggerated story that includes made up details. Samuel wrote down the stories, including one about a frog named Dan'l Webster who could outjump any other frog. On November 18, New York's *The Saturday Press* published the story as "Jim Smiley and His Jumping Frog." For this humorous tall tale, Samuel used his pen name, Mark Twain. Later reprints titled the story, "The Celebrated Jumping Frog of Calaveras County."

The story inspired the Jumping Frog Jubilee® jumping contest. It's been held in Angels Camp, Calaveras County since 1893, as part of the 39th District Agricultural Association county fair.

Over his long career, Twain wrote about the American culture of his times. Famous books by Mark Twain include *Tom Sawyer* and *The Adventures of Huckleberry Finn*.

How Far Can a Bullfrog Jump?

When scientist Henry Astley was studying at Brown University in Rhode Island, he read that in the laboratory, bullfrogs could jump 4 feet 3 inches (1.3 meters).

One day, Henry's supervisor, Tom Roberts, read in the *Guinness Book of World Records* that Rosie the Ribeter had jumped 21 feet 5 and 3/4 inches in her1986 winning triple-jump at the Jumping Frog Jubilee®. That was an average of over seven feet per jump. The statistics didn't match up. Tom and Henry decided to attend the 2013 Jumping Frog Jubilee® to gather more information. They set video cameras up and filmed 3124 frogs jumping, recording over 20 hours of video.

The Jumping Frog Jubilee® provided information on far more frogs than they had ever had in the laboratory. This meant their sample size was larger.

They measured two types of frog jumps.

First were frogs jumped by the professional frog jockeys. They brought their own frogs that had been chosen because they were good jumpers. The frog jockeys knew how to lunge at the frog to make it jump farther. These frogs regularly jumped six to seven feet per jump.

Second were frogs jumped by amateurs, or people who just rented a frog. They knew very little about the frogs or how to encourage the frogs to jump farther. Their frogs only averaged 3.6 feet per jump, similar to those in the lab.

Henry, Tom, and the other scientists concluded that the biggest difference was the frog jockey. The jockeys learned over the years how to encourage the frogs to jump farther. They believe that lunging at the frog makes the frog think a large predator is after them. The jockey's lunge triggers a flight response, the frogs trying to escape the predator.

They also realized that studying or observing animals in a laboratory may be misleading. The laboratory isn't always the best place for an animal to show the behavior that's being studied. For bullfrogs, the Jumping Frog Jubilee® was the best place to show how far they jumped. Or, maybe they jump even farther in the wild when threatened by an actual predator. No one knows.

Sources

Astley, H. C., Abbott, E. M., Azizi, E., Marsh, R. L., and Roberts, T. J. (2013) Chasing maximal performance: A cautionary tale from the celebrated jumping frogs of Calaveras County. Journal of Experimental Biology, 216, 3947-3953. doi:10.1242/jeb.090357

Personal inverviews by telephone with Bill Guzules and Henrietta Guidici, Lee's widow, November 2016.

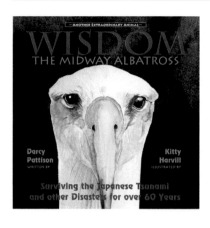

Starred Review-
Publisher's Weekly

2015 NSTA Outstanding
Science Trade Book

2017 NSTA Outstanding
Science Trade Book